The art & artifice of SCIENCE

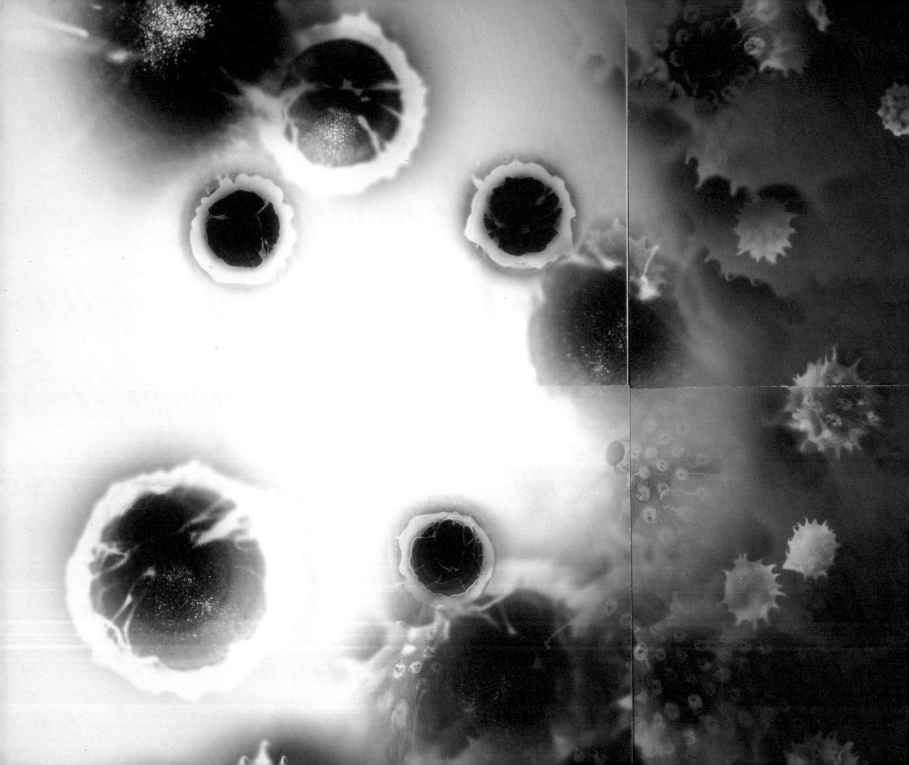

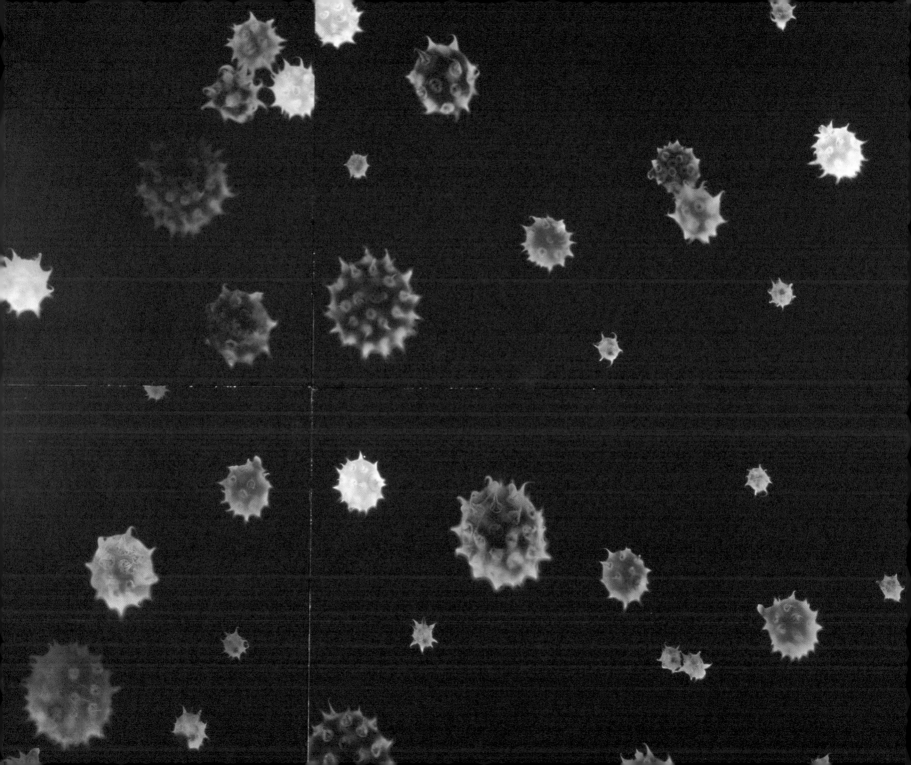

The art & artifice of

SCIENCE

Essays by Laura M. Addison *and* Arif Khan

MUSEUM OF FINE ARTS, SANTA FE

The Art & Artifice of Science is published in conjunction with the exhibition organized by Laura M. Addison and Arif Khan for the New Mexico Museum of Fine Arts, February 9 through May 20, 2007.

A companion exhibition, *Mapping Bodies: The Art & Artifice of Science*, was held at 516 ARTS, Albuquerque, February 10 through March 24, 2007.

Copyright © Museum of Fine Arts, Santa Fe, a division of the New Mexico Department of Cultural Affairs

NEW MEXICO DEPARTMENT OF
CULTURAL AFFAIRS

All rights reserved. No part of this publication may be reproduced in any form or by any electronic or mechanical means, including information storage systems, without written permission of the copyright holders, with the exception of reviewers or scholars, who may quote brief passages in their texts.

Exhibition design: Susan Hyde Holmes and John Tinker
Catalogue design: Susan Hyde Holmes
Photo credits: Craig Weiss (cover, pages 71–73);
        Blair Clark (pages 31–33, 35, 64, 65)

ISBN: 0-9675106-9-4

For inquiries or ordering information:
Museum of Fine Arts
P. O. Box 2087
Santa Fe, NM 87504-2087
505-476-5072
www.mfasantafe.org

COVER: Gail Wight, detail of *J'ai des papillons noirs tous les jours*, 2006, silk, electronics, Plexiglas, 2,800 bug pins. Detail of *Dell'historia naturale di Ferrante Imperato Napolitano*, 1599. Courtesy of Smithsonian Institution Libraries, Washington, D.C.

FRONTISPIECE: Leigh Anne Langwell, detail of *Burst*, 2003, photogram.

CONTENTS

Acknowledgments 9

Foreword *by Marsha C. Bol* 11

PART ONE: *unnatural histories*

Science's Fictions, Art's Wonders 15
by Laura M. Addison

Plates 26

PART TWO: *mapping bodies*

Imagining Science 45
by Arif Khan

Plates 58

Exhibition Checklist 75

acknowledgments

THERE ARE MANY PEOPLE to thank for their support of *The Art & Artifice of Science* and this accompanying catalogue. Doug Ring and Cindy Miscikowski have been steadfast supporters of the Museum of Fine Arts' contemporary art program in the most generous spirit. I am grateful for their beneficence. Friends of Contemporary Art (FOCA) is an essential advocate for the museum's endeavors in contemporary art; they have made possible countless programs, including this exhibition. And Pat Hall, as always, has been an extraordinary champion of our efforts.

The Art & Artifice of Science is the result of a dialogue with my former colleague and guest co-curator Arif Khan. We share a common interest in the intersection of art and science and together we shaped the concept of the show. One could not ask for a more genial collaborator. John Addison and Laura Scholl gave us helpful feedback in the show's development stages.

Many thanks also go to Museum of Fine Arts Director Marsha Bol, Chief Curator Tim Rodgers and Associate Director Mary Jebsen for their interest in and encouragement of this project. Graphic designer Susan Hyde Holmes and exhibition designer John Tinker demonstrated their exceptional talents through the beautiful and nuanced design of the show and catalogue. Preparator extraordinaire Tim Jag expertly installed the show, with the help of Max Friedenberg, Paul Singdahlsen and Jamie Hascall. Our fabulous fabricators—Marvin Valdez, Ron Anaya and Steve Marquez—made the dioramas and other elements of the installation.

Every museum staff member contributes to the development and execution of an exhibition. I want to thank especially Theresa Garcia, Joe Traugott, Ellen Zieselman, Michelle Roberts, Martha Landry, Susan Poorbaugh and summer intern Raven Munsell for their camaraderie and assistance. I am also indebted to public relations manager Steve Cantrell, creative director David Rohr and photographer Blair Clark.

This exhibition provided an opportunity for an unusual collaboration between the Museum of Fine Arts in Santa Fe and 516 ARTS in Albuquerque. Suzanne Sbarge and Andrew John Cecil offered a forum for the companion exhibition *Mapping Bodies: The Art & Artifice of Science,* which fulfilled our desire to curate a show that would bring together not only two disciplines but two cities as well.

Finally, I want to express our appreciation to the exhibition and equipment lenders, as well as several gallerists whose assistance was indispensable: Lynn and George Goldstein, Julie Saul Gallery, Zabriskie Gallery, Bonni Benrubi Gallery, the National Hispanic Cultural Center, the Museum Resources division, Jeremy Lawrence and Rex Jung of the MIND Institute, Hosfelt Gallery, the University of New Mexico's ARTS Lab and Department of Music, Santa Fe Clay and the Andrew Smith Gallery. Most of all, I wish to thank the artists. Their dedication to their work and their sustained engagement with the nexus of art and science are a source of wonder for us all.

Laura M. Addison
Curator of Contemporary Art, Museum of Fine Arts, Santa Fe

foreword

NEW MEXICO HOLDS A UNIQUE PLACE in the history of science and technology. It was in Los Alamos that Robert Oppenheimer and his team of scientists created the atomic bomb. Once built, it was then detonated in the southern New Mexico desert at the Trinity Site in 1945. The group of scientists at Los Alamos formed the nucleus of a "science colony," a group of like-minded people coming together for a common purpose.

Other such "colonies"—art colonies—had already existed in New Mexico since the 1910s, comprised of transplants from the East Coast lured to New Mexico by its cultural richness as well as the broad vistas, the quality of light, the fast-moving storms, the high desert landscape and dry climate. New Mexico's art and science colonies have long coexisted and occasionally even intermingled, establishing this region as a nexus of art and science.

The New Mexico Museum of Fine Arts has had its own role in this dialogue by hosting exhibitions of or collecting works by contemporary artists who address New Mexico's nuclear legacy, among them Tony Price, Woody and Steina Vasulka, Meridel Rubinstein, Patrick Nagatani and Erika Wanenmacher. With the exhibition *The Art & Artifice of Science*, Curator of Contemporary Art Laura Addison and guest co-curator Arif Khan, former curator of the Governor's Gallery, have brought this exploration of art and science into the current century.

Among the many issues in science today are genomic mapping, stem-cell research and alternative uses for nuclear energy. New Mexico's science and technology institutions, including Los Alamos and Sandia National Laboratories, the Santa Fe Institute, the MIND Institute and other research and medical facilities, INTEL and other technology corporations, continue to keep this region at the cutting edge of science. Likewise, the Museum of Fine Arts endeavors to present artistic innovation alongside traditional and modernist works, and the best of regional art alongside international trends—offering a forum for multiple voices and disciplines.

I wish to thank the supporters of this exhibition: Doug Ring and Cindy Miscikowski, Friends of Contemporary Art (FOCA) and the Museum of New Mexico Foundation. And, as always, our gratitude goes to the artists whose work we celebrate in this exhibition, as well as their collaborators in the realm of science.

Marsha C. Bol, Ph.D.
Director, Museum of Fine Arts, Santa Fe

PART ONE | *unnatural histories*

Science's Fictions, Art's Wonders

Laura M. Addison
Curator of Contemporary Art, Museum of Fine Arts

Cabinet of curiosity: Dell'historia naturale di Ferrante Imperato Napolitano, 1599. Courtesy of Smithsonian Institution Libraries, Washington, D.C.

PREVIOUS SPREAD:

Alison Carey, Criptolithus and Eumorphocystis, Ordovician Period, 440–500 Mya, *ambrotype. Courtesy of the artist.*

IN THE 1966 SCIENCE-FICTION MOVIE *Fantastic Voyage,* a heroic crew aboard a submarine miniaturizes itself to microscopic proportions to save a scientist from a blood clot in his brain. The submarine travels through the scientist's body, as if a blood cell, and offers the movie viewer the opportunity to imagine the wonders of the human body from the inside. This and other science-fiction classics captured the imagination of an entire generation awed by the possibilities offered by scientific advances and the prospect of space exploration. *The Art & Artifice of Science* is based upon this same science-fiction mindset engaged by artists whose work speculates about what the world would be like *if.* . . .

THE ARTISTS IN THIS EXHIBITION steer us through our own fantastic voyage based on the model of the natural sciences and biotechnologies once only imaginable in the genre of science fiction and fantasy. By creating works that might be displayed in a faux natural history museum or that would result from a variety of medical imaging techniques, many contemporary artists meld fact and fiction and use the look, feel and visual language of science as their creative impetus. Some create a world inhabited by flora and fauna unlike any we have known before. Other artists collaborate with scientists (real and faux) to create visual imagery that takes us into the world of biotechnology. They challenge us to consider the implications and ethics presented by the Human Genome Project, cloning and genetically modified foods. Many of the artists in *The Art & Artifice of Science* toy with the idea of photographic veracity and scientific objectivity, and provide a critical look at both.

Wondrous worlds

Fact and fantasy often walk hand in hand, particularly in moments of discovery or great change. Indeed, if you consider the European "discovery" of the New World, fantasy was commonplace in the worldview of the new arrivals. As students of Pliny the Elder's encyclopedia of "monstrous races"—among them Cyclops, pygmies, headless men and human/animal hybrids—explorers such as Christopher Columbus expected to locate such wonders in the New World. Columbus reported hearing rumors of an island of Amazons,[i] and sightings of a unicorn, "a feathered monkey that sang like a nightingale, blue men with square heads, cannibals . . . , monsters, ferocious sea animals, and bones of giants"[ii] were all recounted from various sources throughout the sixteenth century. It is no accident that *Wunderkammern* ("wonder chambers"), or curiosity cabinets, the precursors to the modern-day museum,[iii] grew in number with the influx of exotic objects and specimens from the New World during the sixteenth and seventeenth centuries.

The desire (or need) to understand the world around us and to collect it, order it, contain it, classify and categorize it, resulted in fabulous collections of *artificialia* and *naturalia*. Stuffed alligators or rhinoceroses or birds, coral and shells, tiny grains of rice etched with landscapes, scientific instruments, coins, antiquities and paintings all were the source of wonder. This eclectic mix—unified under the principle of displaying "God's ingenuity and fecundity"[iv]—was culled into a single collection which revealed that truth was indeed stranger than fiction. With the Age of Enlightenment in the eighteenth century came scientific inquiry and the dismantling of wonder—and the cabinets of curiosity—in favor of quantifiable data. Wonder, amazement and mystery, as Patrick Mauriés points out, were regarded as "symptoms of ignorance and superstition" associated with "the 'most vulnerable' in society: 'women, the very young, the very old, primitive people, and the uneducated masses.'"[v] Fact and fantasy were divorced.

Fast forward to twenty-first-century Los Angeles, in a Culver City storefront, where one discovers the Museum of

Jurassic Technology, a modern-day cabinet of curiosity. It is the embodiment of the wonder and artifice of science. As Lawrence Weschler writes in *Mr. Wilson's Cabinet of Wonder*, an account of this unique museum, "The visitor to the Museum of Jurassic Technology continually finds himself shimmering between wondering *at* (the marvels of nature) and wondering *whether* (any of this could possibly be true)."[vi] Using the institutional authority of The Museum, the displays leave the viewer confounded. Which exhibitions are real and which are the trickery of a skilled and imaginative curator and exhibition designer? With displays on such topics as the Stink Ant of the Cameroon, the Horn of Mary Davis of Saughall or the science behind the decomposition of dice, the Museum of Jurassic Technology creates a space of ambiguity where one goes to seek knowledge and clarity. But the location of truth is precisely not the point. According to the museum's founder and director, David Wilson, the objective is "to reintegrate people to wonder."[vii] This strategy—of creating ambiguity and wonder—is shared by many of the artists in *The Art & Artifice of Science*.

The artist's greatest asset is imagination, and from imagination springs fictions of all variety. At the same time, the scientist relies on great feats of imagination to come up with hypotheses to prove. Imagination led to the creation of a flying machine, to the discovery that the universe was always expanding, to the breakthrough that humans and chimpanzees are descended from a common ancestor, and to the development of the World Wide Web. Any scientist will attest to the creativity involved in their work, the "fantasizing" that gives birth to a theory—followed by rigorous research to prove or disprove it. The disciplines of art and science, so often represented as polar opposites, in fact share a reliance upon imagination.

Photography's inventiveness

The gray area between fact and fiction has been at the foundation of much contemporary photographic practice, many examples of which appear in *The Art & Artifice of Science*. Since its inception, photography has been considered an instrument of reality, a recorder of life as it *is*. Many a photographer has exploited this assumption to create alternate realities, imagined and imaginary lives, staged scenarios and not quite plausible truths—in other words, to record life as it *could be*. Artifice has been photography's constant companion; as early examples, Hippolyte Bayard feigned his own demise in *Self-portrait as a Drowned Man* (1840) and O. G. Rejlander staged an alleged dream in *The Bachelor's Dream* (ca. 1860).

The Spanish duo Joan Fontcuberta and Pere Formiguera were early, pre-digital-age practitioners of photographic artifice. In the series *Fauna*, they present the journals and findings of Dr. Ameisenhaufen (1895–1955?), purportedly discovered by the two artists in an attic in Glasgow. The *Centaurus neandertalensis* (a horse-humanoid), the *Myodorifera colubercauda* (a squirrel with webbed feet and a snake tail) and the *Improbitas buccaperta* (an alligator-armadillo) are among the species unearthed in Dr. Amiesenhaufen's archives. Are these creatures really outside the realm of possibility? Wouldn't the laws of evolution permit the *possibility* that an *Improbitas buccaperta* existed? If dinosaurs and the dodo once walked the earth, why not a four-legged, armored mammal with a reptilian head and mouth? Surely there are stranger living things that exist in our reality today. If the long-snouted, termite-eating aardvark were to become extinct tomorrow, would someone in the twenty-second century think that this creature was merely a turn-of-the-twenty-first-century practical joke?

Cryptozoology is the study of "hidden animals," or animals that are rumored to exist—most famously, the Loch Ness Monster, Sasquatch (popularly known as Bigfoot) or the jackalope of the American Southwest. Another of these "cryptids," or hidden animals, is the *chupacabras*, a nocturnal creature that kills goats and sucks out all their blood. It has been reportedly sighted throughout Latin America as well as in communities in the Southwest (including New Mexico). While some discount cryptozoology as a pseudoscience and the cryptids as mere fantasy, there are nonetheless individuals who have dedicated their work lives to the study of cryptids, and credible media programs such as *National Geographic* and National Public Radio have devoted printed pages and air time to discussing it. Not unlike the rumors and second-hand accounts that spurred the search for the elusive Plinian races in the New World, the reports and urban legends of cryptid sightings have perpetuated a modern-day version of the search for nature's wonders.

Renowned developmental biologist Pere Alberch once said, "What 'exists' is a small part of what is 'possible.'"[viii] The photographic "discoveries" of Fontcuberta and Formiguera hover between fact and fiction. They reveal layer upon layer of data, like an archaeology of knowledge. Patrick Nagatani has also used archaeology as a metaphor in his series *Excavations*. Ancient ruins in the midst of being uncovered, photographed by Nagatani's alter ego Ryoichi, reveal automobiles embedded within the walls of kivas or underneath sites dating millennia before the invention of the automobile. We recognize the farcical nature of these images—they cannot possibly be true—at the same time that we understand the parody Nagatani is making of the idea that photography is a document of reality. In a twist on Alberch's quote, Nagatani and others exploit the realism of photography to make us see that "what exists (the photograph) is a small part of what is *im*possible."

Resurrection and reinvention

Though we know intuitively that seeing is not believing, photography plays with our willingness, or perhaps our desire, to believe what we see, and to be "reintegrated to wonder"—to quote Museum of Jurassic Technology founder David Wilson. Harri Kallio's resurrection of the dodo started with this desire to see an extinct animal dwell again among the living, at least in the imagination. Working from skeletal remains, anecdotal evidence and drawings, Kallio re-created the dodo and then brought a pair back to its natural habitat, the island of Mauritius, to photograph them in their

Chupacabras (mystery "goatsucker") with prey.
Courtesy of Fortean Picture Library.

FOLLOWING PAGE:

Harri Kallio, dodo reconstruction in progress.
Courtesy of Bonni Benrubi Gallery, New York.

Science's Fictions, Art's Wonders

environs. Alison Carey engaged in a similar scholarly re-creation of natural history in her series *Organic Remains of a Former World*. She consulted with paleontologists and studied scientific illustrations of Paleozoic fossils, dating from 230 to 570 million years ago, to create a model of ocean life that she then documented with modern ambrotypes.

Kallio's and Carey's re-creations of the past are based on fact—at least as far as has been determined by scientists to date—yet they maintain an element of falseness. The image of a dodo captured in full color is anachronistic; it became extinct in the mid-seventeenth century, long before the advent of photography. And to the layperson, the underwater plants and animals of Carey's photographs are so dissimilar to the oceanic life we know today that they seem otherworldly. This shakes all confidence in our sense of chronology and our understanding of "truth." Both Kallio and Carey have made photography not only a means of documentation of a moment in time, but also an instrument of resurrecting the past.

Art can also be a forecast of a possible future. Each generation has new scientific advances to grapple with, accompanied by their own ethical ambiguities. Since the cloning of Dolly the sheep in 1996, science revealed new issues for artists to explore: biotechnology, cloning, genetic engineering. In the post-Dolly reality, commercial enterprises such as Genetic Savings & Clone, Inc. have emerged that allow you to save your favorite pet's DNA for future cloning, for a mere $32,000.[ix] Creatures such as Rebekah Bogard's mutant pig alien or Laurie Hogin's strangely human monkey seem less inconceivable now than they did in the golden age of science fiction.

Photographs are another sort of replicant, a cloned positive based on an original negative. Both Daniel Lee and Christine Chin create hybrids that may have been imaginary, but with today's technology... could they be in the realm of "the possible"? Daniel Lee's human/animal hybrids such as *Leopard Spirit* or *Pig King* are so seamlessly morphed that the latent animal attributes that come through in his sitters' portraits are a reminder that humans and chimpanzees share nearly 99% of their genetic make-up. Christine Chin's human/vegetable hybrids are at once humorous and unsettling. A plant and an animal could never hybridize as these images suggest; instead the chatty tomatoes, the "pota-toes" and the garlic nose are an ironic commentary on genetically modified foods, an example of the debate between the merits and potential dangers of technology.

Institutional knowledge

The Museum and The Laboratory and The University are institutions that construct and disseminate knowledge, and thus present themselves as exclusive guardians of truth. The design of a space—be it a natural history museum or a laboratory—contributes to the institution's ability to convince us of this guardianship. The dwarfing neoclassical façade of the American Museum of Natural History, with an equestrian statue of pres-

Alison Carey, Phyagmoceras & Pseudocrinites, Silurian Period, 400–400 Mya, *2005, ambrotype. Courtesy of the artist. Exhibited at 516 ARTS, Albuquerque.*

ident and naturalist Theodore Roosevelt situated before the entrance, gives the visitor a preview of the enlightenment he or she will soon experience. Inside, the four-story maze of galleries organizes the natural world according to taxonomic categories and regions. Peoples from various continents find their places among the displays as well. The quintessential display in the museum is, of course, the diorama, a term coined, interestingly enough, by inventor of the daguerreotype, Louis Jacques Mandé Daguerre. A diorama presents in great detail a vignette from nature, complete with painted landscape in the background and trees, rocks, grasses or other elements of the animals' habitat in the foreground—all with a false perspective to give the illusion of depth. The stuffed animal at the focal point of the diorama is caught mid-action doing whatever is characteristic of that animal. Like a photograph, the diorama is an approximation of the real, and in its amazing craftsmanship, is as much an object of wonder as is the natural world it depicts.

The sterile, high-tech, white-walled space of a laboratory lacks the embellishments of a museum, but it still conveys certain ideas to the visitor: science is pure, untainted by human emotion or error, and fact-based (machines don't lie). We have confidence in the conclusions that come from these spaces, and the machinery and interior design reinforce this trust. Just as a natural history museum demonstrates man's imposed order on the natural world, the laboratory shows man's mastery of knowledge and technology in the name of progress. Machines take center stage; drawers and cabinets order the space—reminiscent of the manner in which cabinets of curiosity ordered their specimens. If the diorama is a window upon the natural world, the laboratory is a window upon the world on a microscopic scale. Microscopes, MRI machines and other instruments of seeing show us internal workings of living organisms and seek explanations and solutions. The technology that allows us to see inside the body is as much an object of wonder as is the body itself.

GAIL WIGHT'S PLEXIGLAS MICROSCOPE, *Ghost*, is an icon of the instruments of seeing that have allowed science to disentangle the mysteries of life. On one level, the insect that pulsates with light on the sacrificial altar of Wight's microscope is an homage to all of the insect souls lost in the name of science. On another level, the transparency of the microscope clone is a metaphor for the elusiveness of truth, fact and knowledge. "Wherever science falls out of the empirical and lapses into the poetic, or simply ceases to make sense," she says, "this is the place where art begins for me."[x] In other words, Wight's art, like this exhibition, is meant to "reintegrate people to wonder" and to allow fact and fantasy to walk hand in hand once again.

i *The Diario of Christopher Columbus's First Voyage to America, 1492–1493*, abstracted by Fray Bartolomé de las Casas (Norman and London: University of Oklahoma Press, 1989), 331.

ii Abby Wettan Kleinbaum, *The War Against the Amazons* (New York: New Press, 1983), 111.

iii Celeste Olalquiaga, "Object Lesson / Transitional Object," *Cabinet* 20 (winter 2005/2006): 7.

iv Stephen T. Asma, *Stuffed Animals and Pickled Heads: The Culture and Evolution of Natural History Museums* (New York: Oxford University Press, 2001), 43–44.

v Cited in Patrick Mauriés, *Cabinets of Curiosities* (London: Thames & Hudson, 2002), 193.

vi Lawrence Weschler, *Mr. Wilson's Cabinet of Wonder* (New York: Vintage Books, 1995), 60.

vii Ibid.

viii Pere Alberch's quote is the opening page of Joan Fontcuberta and Pere Formiguera's website portfolio of the *Fauna* series: http://zonezero.com/exposiciones/fotografos/fontcuberta

ix During research for this essay, the services of Genetic Savings & Clone were offered on the website www.savingsandclone.com. By press time, this company was closed and the home page of its website directed visitors to ViaGen, with the stated caveat: "Note that ViaGen has no plans to provide commercial cat or dog cloning services."

x Gail Wight, "Artist Statement," http://vv.arts.ucla.edu/terminals/t1/ucsc/wight/wight.html

What "exists" is a small part of what is "possible."

—Pere Alberch

Rebekah Bogard

Kisses, 2005

Earthenware, underglaze, glaze, metal rod

27 x 32 x 23 inches

Courtesy of Lynn Marchand Goldstein and George Goldstein

Alison Carey

Stethacanthus, Pennsylvanian Period, 280–310 Mya, 2005
Ambrotype
9 x 23 inches
Courtesy of the artist

Christine Chin

Hybrid Tomatoes, 2003–2005
Digital inkjet print
20 x 30 inches
Courtesy of the artist

Joan Fontcuberta and Pere Formiguera

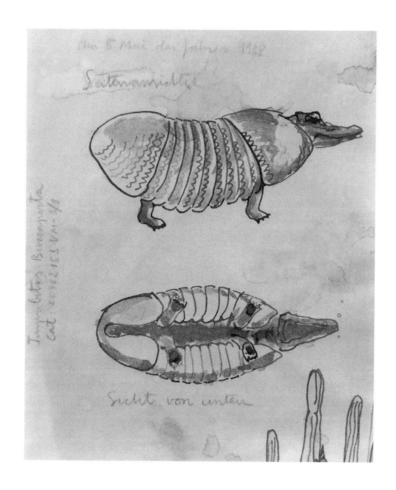

From *Improbitas buccaperta*
Ink drawing
6 x 5 inches
Courtesy of Zabriskie Gallery, New York

OPPOSITE:

From *Improbitas buccaperta*
Gelatin silver print
20 x 16 inches
Courtesy of Zabriskie Gallery, New York

Laurie Hogin

Still Life with Broken Fruit, 2002

Oil on panel

21 ½ x 17 ¼ inches

Courtesy of Lynn Marchand Goldstein and George Goldstein

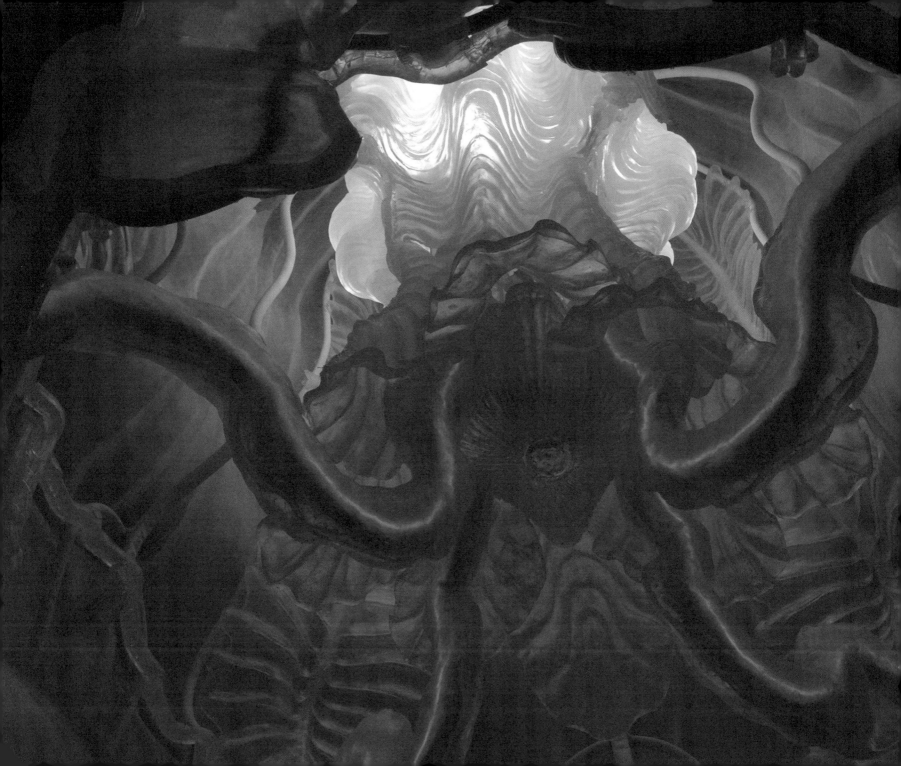

Timothy Horn

Stheno, 2006

Silicone rubber, copper tubing, fiber optics, lighting fixtures

60 x 47 inches diameter

Courtesy of Hosfelt Gallery, New York

Harri Kallio

Mare Longue Reservoir #1, Mauritius, 2002
Digital chromogenic print
29 ½ x 35 inches
Courtesy of Bonni Benrubi Gallery, New York

OPPOSITE:

Riviere des Anguilles #7, Mauritius, 2002
Digital chromogenic print
29 ½ x 35 inches
Courtesy of Bonni Benrubi Gallery, New York

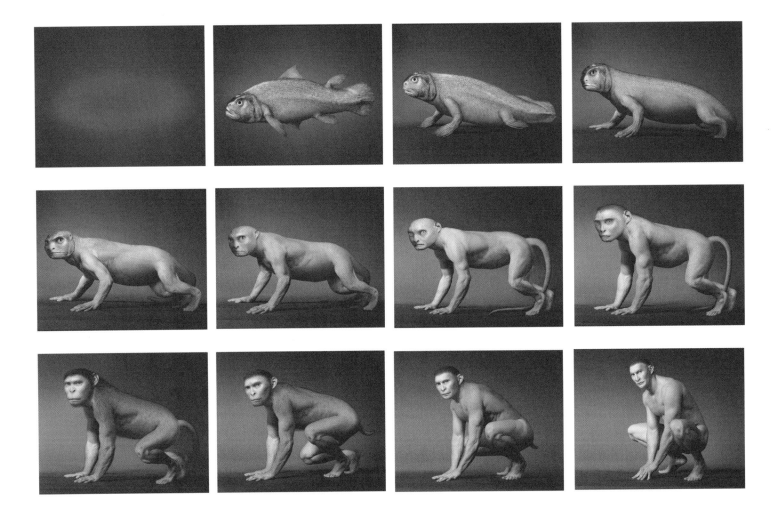

Daniel Lee

Juror No. 6 (Leopard Spirit), 1994
Digital inkjet print
53 x 35 inches
Collection of the Museum of Fine Arts,
Gift of Marvin and Marlene Maslow, 2003

OPPOSITE:
Stills from *Origin*, 1999–2003
Digital animation
5 minutes
Courtesy of the artist

PART ONE: UNNATURAL HISTORIES

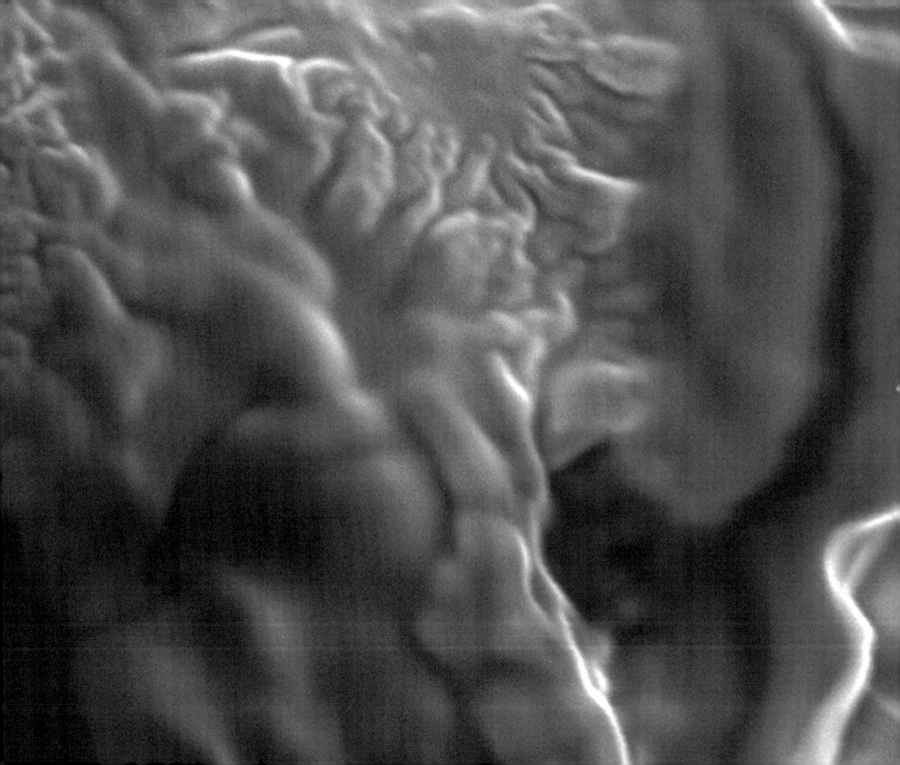

PART TWO | *mapping bodies*

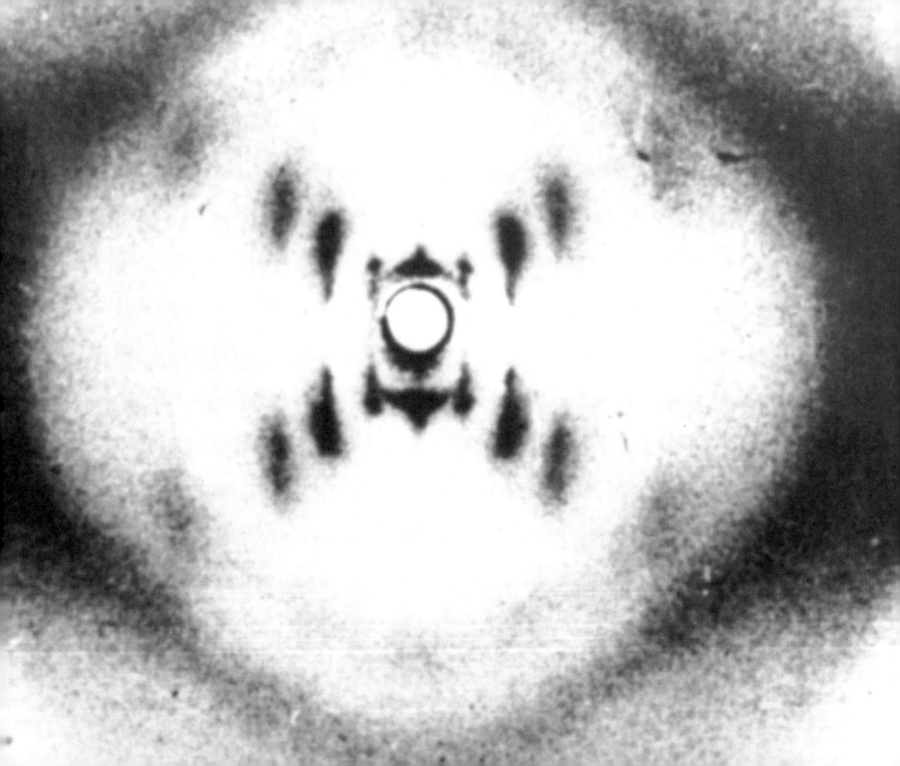

Imagining Science

Arif Khan
Guest Co-curator, The Art & Artifice of Science
Gallery Director, Tamarind Institute

LEFT:
First photograph of DNA B-form, by Rosalind Franklin, 1952. Courtesy of Cold Spring Harbour Laboratory Archives.

PREVIOUS SPREAD:
Justine Cooper, video still from Scynescape, *2000/2006, scanning electron microscopy animation. Courtesy of the artist.*

THE EXPLORATION OF WHAT IT MEANS to "be human" is a hallmark of contemporary art practice and theory. We are a society obsessed with identity, seeking its origin, definition and construction. The latter half of the twentieth century witnessed artists investigating notions of national, ethnic and sexual identity that, in their view, are shaped and defined by external societal forces. Citing erroneous medical explanations for the place of women in society or for the superiority of one race over another (eugenics), some of these artists argued that the sciences served to enforce these social constructs. Recent developments and advances in the sciences, specifically scientific imaging, have allowed artists to reconsider the relationship between the visual arts, science and society. The work presented in *The Art & Artifice of Science* questions the limitations of the social construction of identity and suggests that we may need to incorporate a biological perspective into our understanding.

Genetic identity

Launched in 1990, the Human Genome Project promised to reveal the genetic blueprint that tells us who we are. The "gene" is represented in popular culture as the source of individual identity and the agent of cultural formation. During the 1990s, the news media featured headlines that remarked on the discoveries uncovered by genetic researchers. Many of these concerned genes that are responsible for various diseases, such as leukemia and multiple sclerosis. However, the media also touted research that provided genetic explanations for conditions previously considered environmentally or socially constructed. There are pleasure-seeking genes, divorce genes, violence genes, genes for addiction, sexuality and genius. These popular depictions portray the gene as powerful, deterministic and central to the understanding of everyday behavior.

In 1996 Gary Schneider was commissioned to make photographs in response to some of the revolutionary discoveries that were emerging from the Human Genome Project. Schneider consulted with doctors and geneticists, examined diagnostic and forensic photographs, and created x-rays, photograms and micrographs of various parts of his own body. The resulting *Genetic Self-Portrait* explores the fields of biology, genetics and forensic science and how those disciplines affect personal notions of privacy and narcissism.

Increasing popular acceptance of genetic explanations for the creation of identity and the proliferation of genetic imagery substantiate the highly publicized research in the science of genetics. Such research occurs in a specific cultural context and is fueled by our creation of social policy and practice based on notions of heredity and "natural" ability. The "gene" as a deterministic agent for human behavior and a basis for social relations promises the possibility of even greater certainty, order, predictability and control.

Schneider's *Genetic Self-Portrait* was produced at a time when the science of genetics moved from the laboratory to mass culture, from professional journals to popular magazines and television. This shift transformed the definition of the "gene." Instead of a piece of hereditary information, it has become the key to human relationships, the essence of personal identity and the source of social difference.

Popularization, proliferation, recontextualization

The convergence of art and science has been a gradual process prompted by contemporary artists' growing interest in the processes and products of science and the popularization of science in the media. Resulting from this increased popularity, a number of collaborative projects arose from the intersection of art, technology and science, including some that appear in *The Art & Artifice of Science*.

What is the attraction of scientific images for artists and viewers and why have we recently become increasingly drawn to such images? Many of these images, of course, have a quality that is undeniably "aesthetic" and are "beautiful" whether they are understood in their particular contexts or viewed as things apart. Scientists often express a sense of wonder and exhilaration at the forms they study: the perfect numerical order of a Fibonacci sequence in the whorls of a shell, the delicate asymmetry of cells, the scratchy traces of subatomic particles.

It is possible to view these objects as lovely manifestations of a beautiful impersonal world, though it should not be forgotten that scientific images can also be highly evocative. Featuring actual human corpses encapsulated in polymer and displayed in a variety of ordinary and fantastic poses, the blockbuster exhibition *BODY WORLDS* has crisscrossed the globe. It draws vast crowds eager to see objects which may be commonplace for many medical scientists, but which are no longer visible in our current, sanitized lives. For most artists, and indeed most of the public, the world is not impersonal; displayed scientific images will be viewed subjectively, in relation to the viewer's own feelings and experience, just as art objects are viewed in museums. With scientific images, we are perhaps confronted, more vividly than with conventional artworks, by a sense of our own impermanence and fragility, a sense in which our consciousness of identity and scale is displaced.

Scientific imaging technologies and their resulting images and methodologies provide material for artist and scientist to begin collaboration. Scientists create and use visual representations that they believe convey an objective account of "reality" or "truth" that conforms to the scientific method. A growing number of visual and sound artists are attracted to these images of science and pursue active collaboration with scientists and scientific institutions.

This is the case with a collaborative project on the visualization and sonification of the kidney by a consortium of University of New Mexico and University of Hawaii researchers.[i] Bringing together music, visuals, computer science and biology, the group sought to clarify how the nephron of the kidney functions. The result is a project that conveys visual and aural beauty, as much as it serves as an educational vehicle for medical students. Using the visual metaphor of the landscape and duct systems to represent the inner workings of the body—complete with a mountain range, skyline and piping—the viewer navigates this virtual renal reality. The algorithmically generated music, by Panaiotis, propels you through the body at different tempos to correspond to the particular molecular particle flow. Displayed in the context of an art museum, one appreciates it as a work of visual and sound art. Seen at a scientific conference, it is understood on completely different terms.

The visual material that science has until relatively recently kept to itself—whether body parts or MRIs, pictures of the stars from the Hubble Telescope or traces of subatomic particles—is now becoming much more available. Our growing

Young visitors at BODY WORLDS *in Los Angeles, 2005. Courtesy of* Gunther von Hagens' BODY WORLDS *and the Institute for Plastination.*

curiosity says much about our desire to view the world and ourselves anew through the filter of science.

When one considers scientific images as a distinct category, it is apparent that scientific representations have always been, and still are, influenced by aesthetic concerns. This is most obviously the case with the Naturalists of the nineteenth century, who often were both artist and scientist, experienced in both seeing and drawing. For contemporary scientists, visual skills are less of a requisite (though exceptions exist, especially in the medical and biological sciences). Contemporary science visualizes its data using ever-changing technology such as electron microscopy, color-enhanced MRI images, virtual reality technology and the various software applications designed to study the gene.

The images produced by such technologies are often not actual representations of scientific data. For example the color, scale, definition and visual user-interface that characterize them are frequently "artificially" added to the raw data. This is most often accomplished at computer terminals in a manner that suggests that they are not purely scientific choices. Through these sites of scientific image production, one can see that the strict divide between a mechanical representation of "reality" and an aesthetic construction of that same "reality" often becomes blurred. One such scientific application is the database visualization tool LDRDView™, developed by Sandia National Laboratories. Research scientists utilize this tool to organize and disseminate the ever-increasing amount of data produced by the Human Genome Project.

LDRDView™ assists the user in discovering unexpected relationships in extremely large collections of data. It uses the terrain or landscape metaphor, condensing groups of similar data sets by physically situating them close to each other in the landscape. The ability of LDRDView™ to visualize vast amounts of scientific data marks a shift from previous visualization technologies. The history of scientific visualization usually entailed the altering or destruction of the specimen being studied. For example, the x-ray photograph of DNA produced by Rosalind Franklin in 1952, which assisted in discovering the double helix structure of DNA, is not an image of a cell or any other living object. It is an image of a crystalline structure composed from various cell extracts, a product of extensive preparations and manipulations of the contents of a vast number of cells, which were in turn destroyed to produce the resulting image. For the research scientist the function of visualization tools such as LDRDView™ is not simply to enhance looking, or even to validate the tangibility of what is being imaged, but to employ the act of looking as an examination for future action upon a patient or specimen.

Mapping bodies

To make oneself the subject of a scientific study is potentially to see one's body as neutral site, where physiological functions can be both disclosed and affected by invasive and non-invasive technologies alike. The work of video and installation artist Justine Cooper questions the veracity of scientific visualization

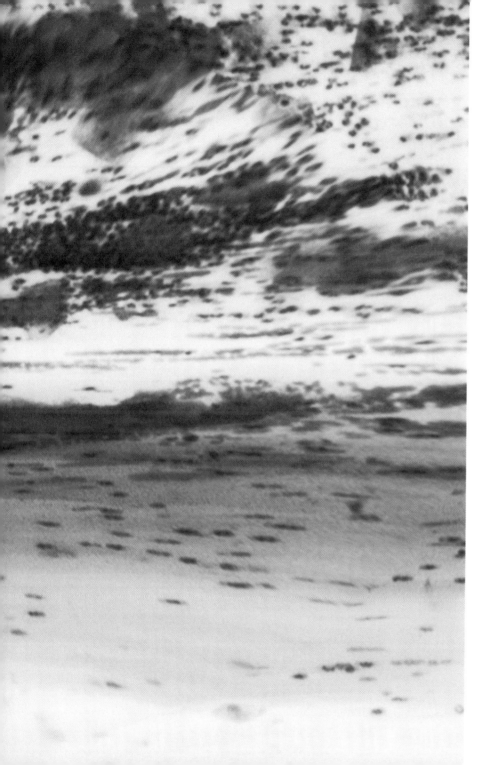

technologies and the resulting scientific image by presenting the viewer with heavily manipulated microscopic imagery of the artist's own body. In *Scynescape* the artist uses scanning electron microscopy (SEM) to create images of molds made from her skin and hair. The hair has been coated in gold to give the electron beam in the microscope something to conduct through in order to render an image, as none of these imaging techniques uses light. The resulting images are projected in an environment that is immersive visually and aurally. The viewer feels as if he or she were navigating through an imagined microscopic world, a topography that is both familiar and strange at the same time.

Threshold, the debut work by the art collaborative M-M-M (mind-meets-matter), explores the relationship between external experiences through the skin (touch) and internal mind (thought). The project uses as its foundation research that looks at skin, our largest organ, as a sort of "external brain."[ii] Using a touch screen, the artists built an interactive interface for the viewers to trigger video images, supplied by research professor Rex Jung of Albuquerque's MIND Institute,[iii] placing their hands on a clear screen devoid of images. The "touch" activates images projected on a wall that in turn "touch" the eye and the mind. According to the artists, "Through our sense of touch we learn, discern, feel and chronicle the outside world; in turn we sweat, we feel pain, pleasure, are aroused, we shiver,

Justine Cooper, video still from Moist, *2002, light microscopy animation. Courtesy of the artist. Exhibited at 516 ARTS, Albuquerque.*

Imagining Science

we redden, become ashen all in response to thoughts we have. The cycle will attempt to make one aware of emotional and physical states of being and the connections between touch and consciousness."[iv]

Cooper's *Scynescape* and M-M-M's *Threshold* present the viewer with an imagined reformulation of the relationship between the body's exterior and interior. The body is viewed as a landscape through which the viewer can travel and ponder the corresponding detachment and personal relationship scientific images produce. The immersive and interactive imagery of these artworks is assimilated within the mind and then fixed into an intimately private, yet simultaneously cultural and social identity. As viewers, we recognize the body imagery as generically human and wonder at the degree to which the external physical body relates to social identity, and the degree to which the individual ("me") corresponds to the human ("us").

These issues, as well as scientific veracity, are also explored in Leigh Anne Langwell's photograms. With a background in biological and medical imaging, Langwell creates her imagery in the darkroom without a camera or negative, laying her own latex sculptures on photographic paper and exposing it to light pulses. The resulting shadowy images seem to offer a view into the inner workings of the human body—or do they document the traces of light produced by celestial bodies? A microscopic interpretation of the imagery leads to the question, Is this what my cells really look like? Are these intimate views of a healthy or a sick individual? The spaces offered in these images act as a conceptual landscape, where notions of abstract and literal, microscopic and macroscopic identity are explored. The ambiguity offered by Langwell's work is a reminder that photography has never managed to fulfill its early aspiration to represent objective reality, but that it has instead always consisted of a compelling but uneasy amalgam of fact and fiction.

Reaction and recognition

Instruments of visualization, like the telescope, microscope and, more recently, MRIs, marked a shift from the use of the naked eye to the manipulation of machines for seeing. An important consequence of this change was that the trust traditionally placed in the observer's senses is now placed in these instruments and the claims made about exactly what could be seen with them. Simultaneously compelling and disconcerting, the imagery produced by these tools forced the public to reconsider long-held notions of the body and its relationship to public and private identity. On a more basic level, these images prompted the question, Is this really what I/we look like? Does this technology change how I see myself or how others see me? This skepticism about whether, and how, to trust the mechanical extension of sight is nothing new and was extended to another technical innovation—the camera—more than a century ago.

Since photography's inception, it has been linked to the practices of the scientific community. On June 15, 1839, writing in his capacity as the Secretary of the Academy of Sciences, François Arago recommended that the French government

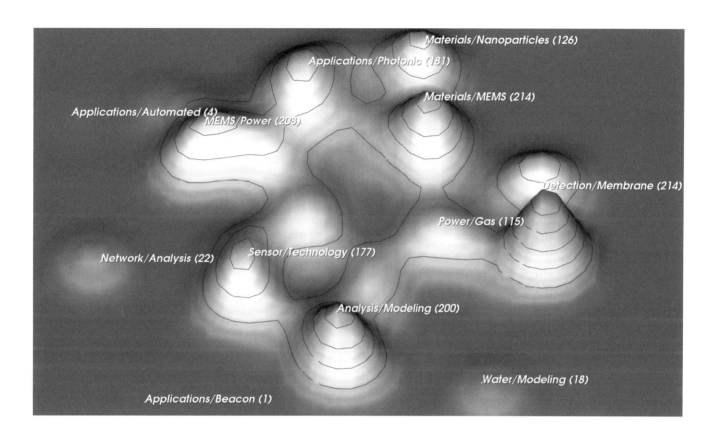

LDRDView™ Visualization Application. Courtesy of Sandia National Laboratories. Sandia is a multiprogram laboratory operated by Sandia Corporation, a Lockheed Martin Company, for the United States Department of Energy's National Nuclear Security Administration under contract DE-AC04-94AL8500.

Imagining Science

should purchase the patent on the invention of photography from Louis Jacques Mandé Daguerre and make it available to the public. Linking photography to the sciences, Arago described the process as follows: "In fact to the traveler, to the archeologist and also to the naturalist, the apparatus of Daguerre will become an object of continual and indispensable use. It will enable them to note what they see, without having recourse to the hand of another."[v] Proceeding from this announcement, the development of photographic technology has continued to refine methods for capturing and recording ever more hidden natural structures and patterns, as well as objects in motion and changing through time. Photography profoundly altered the process of obtaining visual information, representing both a new type of permanent record of observation itself and a new medium through which scientific data could be communicated. Most importantly, photography changed how science was practiced and how it was publicly received. The pre-photographic world of science was a dramatically different place.

Patrick Nagatani's series *Chromatherapy,* which consists of elaborate staged photo-dramas, shuns the new instruments of visualization and looks back to an "alternative" healing practice. With roots in ancient Egypt and China, chromatherapy involves the belief that shining the rays from colored lamps on afflicted areas can cure diseased organs of the body. Nagatani seeks to create "cinematic narrative images that are medical charades of chromatherapy." For this exhibition he has installed an abandoned chromatherapy facility in the Museum, displacing the traditional physical boundaries placed on art and medicine and further altering the contexts through which the viewer normally approaches a medical object. The chromatherapy treatments depicted in Nagatani's photographs put in question how the external environment impacts the internal body. How do chromatherapy lamps exposed to the exterior body (the skin) affect the internal health of the body? What, if any, are the unseen forces positively or negatively impacting the body? And how do we define health and sickness without the benefit of these instruments of seeing? One hopes that the desired healing properties of chromatherapy are at work. However there is a pervading sense of skepticism without the quantitative and pictorial measures of health to which we have become accustomed.

CONTEMPORARY SCIENCE HAS TRANSFORMED our view of the body as never before, dislocating and fragmenting our sense of personal identity, of physical boundaries, scale and time. Like artists, scientists transmute and re-create. Like scientists, artists have begun to appropriate a quasi-scientific neutrality and detachment, utilizing scientific instruments and technology and, in so doing, they question and unsettle the relationship between body and personal identity. Some of the most difficult subject matter has been transmuted through photography, video and film. As spectators, we stand sufficiently distanced by science to look into another's body processes and see real bodies objectified. In the hands of artists, this neutral science becomes art that provokes fundamental questions of identity and, ultimately, of being.

i   This project, called Project TOUCH (Telehealth Outreach for Unified Community Health), began as a medical education study. Flatland, "an open source visualization and virtual reality application development tool created at the University of New Mexico," was used to develop a means of teaching students about the way in which the kidney nephron works, using visuals and music. Beyond its original intended purpose, this virtual reality (VR) simulation is also indicative of the aesthetic choices made in medical imaging. See V. M. Vergara, Panaiotis, T. Eyring, J. Greenfield, K. L. Summers, T. P. Caudell, *Flatland Sound Services Design Supports Virtual Medical Training Simulations,* MMVR 14 Proceedings, IOC Press, The Netherlands, 2006. Also, Panaiotis, V. Vergara, A. Sherstyuk, K. Kihmm, S. M. Saiki, D. Alverson, T. P. Caudell, *Algorithmically Generated Music Enhances VR Nephron Simulation,* MMVR 14 Proceedings, IOC Press, The Netherlands, 2006.

ii Bernadette Healy, M.D., "Skin Deep," *U. S. News & World Report,* http://www.usnews.com/usnews/health/articles/051114/14skin.htm.

iii "The MIND Institute is a consortium of internationally recognized neuroimaging scientists and medical research institutions who have joined together to pursue crucial questions about the mechanisms and treatment of mental illness.... Many different brain imaging techniques can be used to study the mind and brain, such as structural magnetic resonance (sMR), functional magnetic resonance (fMR), magnetoencephalography (MEG), magnetic resonance spectroscopy (MRS), and optical imaging. Each technology alone has limitations. Therefore, the MIND Institute has launched an initiative to combine these multiple tools to create a 'virtual' image of the workings of the mind and brain." www.themindinstitute.org.

iv Min Kim Park, Masumi Shibata and Mary Tsiongas (M-M-M Collaborative), artist statement, December 2006.

v   François Arago, in his report to the Chamber of Deputies on June 15, 1839, as reproduced in Vicki Goldberg, ed., *Photography in Print: Writings from 1816 to the Present* (New York: Simon & Schuster, 1981), 32.

All my originality consists in bringing fantastic beings to life by making them plausible, and, as much as possible, in putting the logic of the visible at the service of the invisible.

—Odilon Redon

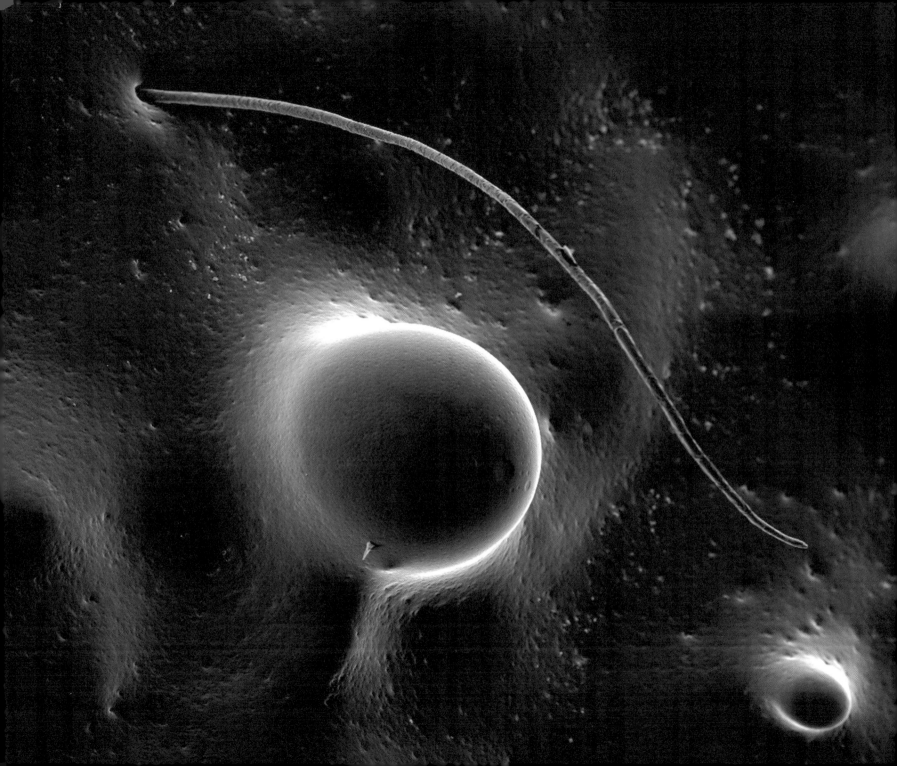

Justine Cooper

Trap (Self-portrait), 1998
14 x 12 x 12 inches
Plexiglas and film
Courtesy of the artist

OPPOSITE:
Still from *Scynescape*, 2000/2006
Scanning electron microscopy animations
Dimensions variable
Courtesy of the artist

Leigh Anne Langwell

Maculae, 2002
Photogram
39 ¼ x 31 ½ inches
Courtesy of the artist

OPPOSITE:
Drift, 2002
Photogram
19 ⅝ x 31 ½ inches
Courtesy of the artist

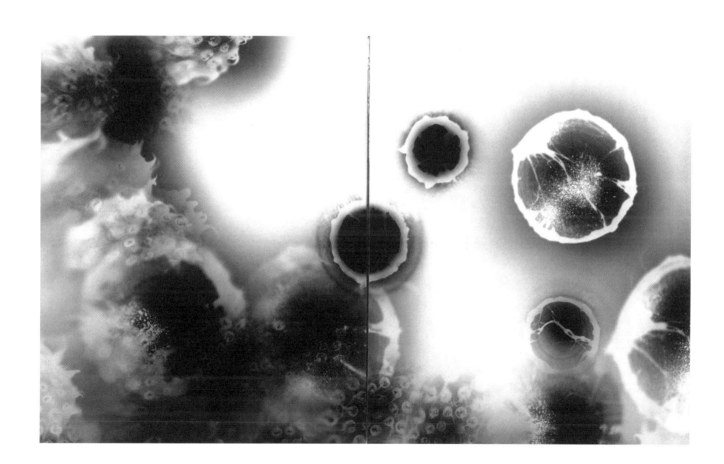

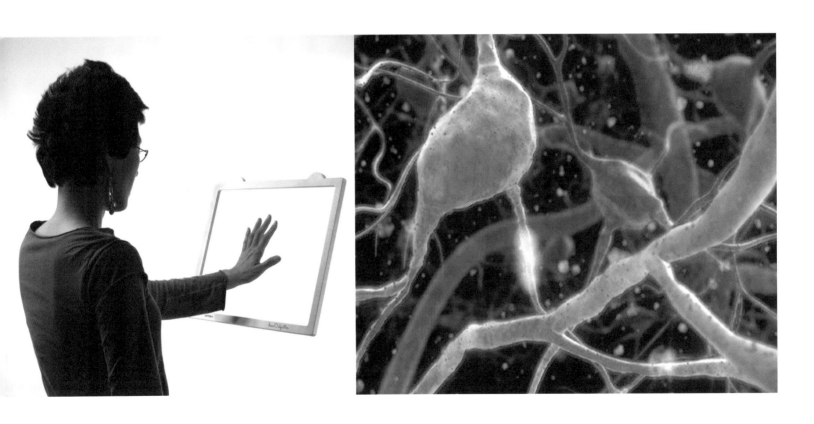

The M-M-M Collaborative

Threshold, 2006

Interactive touch screen, projection

Dimensions variable

Courtesy of the artists and the MIND Institute, Albuquerque

Patrick Ryoichi Nagatani

Tonation in Color Charged H$_2$O, 2004
Chromogenic print, Type C, Ilfoflex 2000
10 x 18 inches
Courtesy of Andrew Smith Gallery, Santa Fe

Sliding Filament Theory—Micro Chromatic Healing, 2006
Chromogenic print, Type C, Ilfoflex 2000
10 x 18 inches
Courtesy of Andrew Smith Gallery, Santa Fe

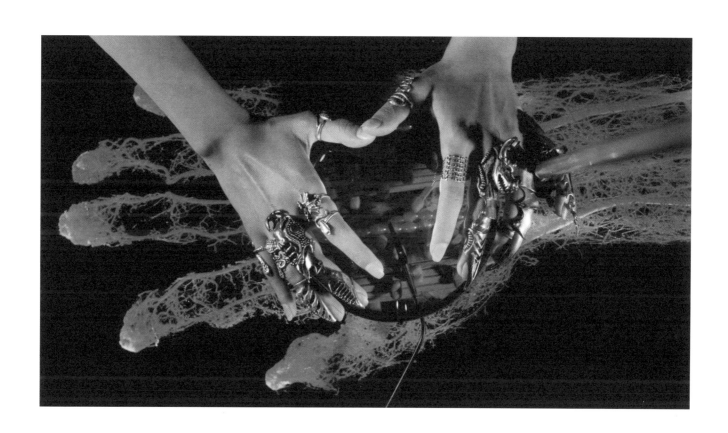

Panaiotis

with Dale C. Alverson, Thomas P. Caudell, Chris Davis, James Holten III, Stanley M. Saiki, Andrei Sherstyuk, Ken Summers and Victor M. Vergara

The Waters of Life: A Reified Voyage into the Kidney, 2004–2006
Computer simulation, modeling and interactive virtual reality
Dimensions variable
Courtesy of University of New Mexico Health Sciences Center for Telehealth and the Center for High Performance Computing Visualization Laboratory

Gary Schneider

Hands, 1997
Pigmented ink on paper
56 x 44 inches
Courtesy of the artist and Julie Saul Gallery, New York
Exhibited at 516 ARTS, Albuquerque

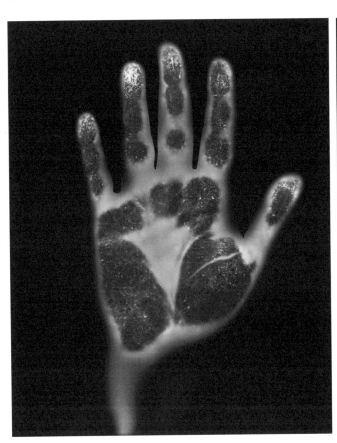
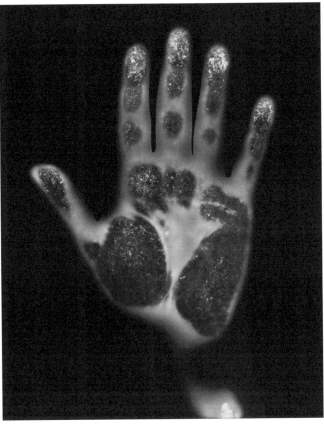

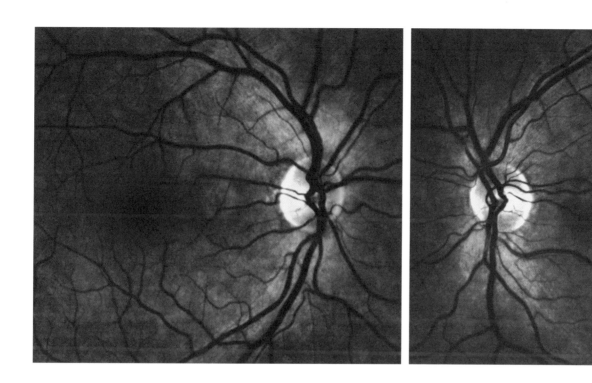

Retinas, 1998

Gelatin silver prints (diptych)

19½ x 20½ inches each

Courtesy of the artist and Julie Saul Gallery, New York

Gail Wight

Ghost, 2004
Plexiglas, electronics
12 x 7 x 8 inches
Courtesy of the artist

FOLLOWING SPREAD:
Details of *J'ai des papillons noirs tous les jours,* 2006
Each of twenty-eight measures 5½ x 5½ x 3½ inches
Silk, electronics, Plexiglas, 2,800 bug pins
Courtesy of the artist

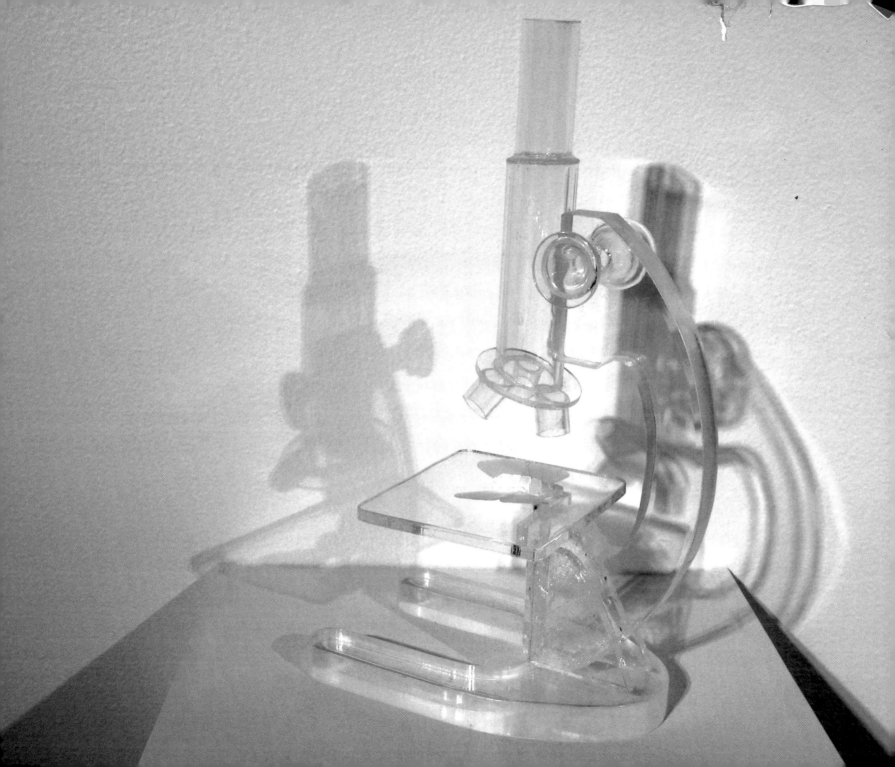

exhibition checklist

Rebekah Bogard

Kisses, 2005

Earthenware, underglaze, glaze, metal rod

27 x 32 x 23 inches

Courtesy of Lynn Marchand Goldstein and George Goldstein

Alison Carey

Diorama: ceramic, contemporary ambrotypes from the series
Organic Remains of a Former World:
a) *Criptolithus and Eumorphocystis, Ordovician Period, 440–500 Mya*
b) *Stethacanthus, Pennsylvanian Period, 280–310 Mya*

Diorama: 42 x 60 x 48 inches; each ambrotype 9 x 23 inches

Courtesy of the artist

Christine Chin

Puckered Peppers (from the series *Vegetable Human Hybrids*), 2003–2005

Digital inkjet print

20 x 32 inches

Courtesy of the artist

Christine Chin

Potatoes (from the series *Vegetable Human Hybrids*), 2003–2005

Digital inkjet print

20 x 28½ inches

Courtesy of the artist

Christine Chin

Hybrid Tomatoes (from the series *Vegetable Human Hybrids*), 2003–2005

Digital inkjet print

20 x 30 inches

Courtesy of the artist

Christine Chin

Garlic (from the series *Vegetable Human Hybrids*), 2003–2005

Digital inkjet print

28 x 23½ inches

Courtesy of the artist

Justine Cooper

Scynescape, 2000/2006

Scanning electron microscopy animations

Dimensions variable

Courtesy of the artist

Justine Cooper

Trap (Self-portrait), 1998

Plexiglas and film

14 x 12 x 12 inches

Courtesy of the artist

Joan Fontcuberta and Pere Formiguera

Improbitas buccaperta (from the series *Fauna*)

Three gelatin silver prints, 20 x 16 inches each;

two ink drawings, 8 x 5¼ inches and 6 x 5 inches;

two hand-written notes, 6 x 8¼ inches each

Courtesy of Zabriskie Gallery, New York

Laurie Hogin

Still Life with Broken Fruit, 2002

Oil on panel

21½ x 17¼ inches

Courtesy of Lynn Marchand Goldstein and George Goldstein

Timothy Horn

Stheno, 2006

Silicone rubber, copper tubing, fiber optics, lighting fixtures

60 x 47 inches diameter

Courtesy of Hosfelt Gallery, New York

Harri Kallio

Dodo Reconstruction, 2006

Aluminum, steel, feathers, silicon and latex rubber, fiberglass, polyurethane foam, glass

44 x 14 x 30 inches

Courtesy of Bonni Benrubi Gallery, New York

Harri Kallio

Mare Longue Reservoir #1, Mauritius, 2002

Digital chromogenic print

29½ x 35 inches

Courtesy of Bonni Benrubi Gallery, New York

Harri Kallio

Riviere des Anguilles #7, Mauritius, 2002

Digital chromogenic print

29½ x 35 inches

Courtesy of Bonni Benrubi Gallery, New York

Leigh Anne Langwell

Ejecta, 2003

Photogram

31½ x 19⅝ inches

Courtesy of the artist

Leigh Anne Langwell

Maculae, 2002

Photogram

39¼ x 31½ inches

Courtesy of the artist

Leigh Anne Langwell

Drift, 2002

Photogram

19⅝ x 31½ inches

Courtesy of the artist

Leigh Anne Langwell
Burst, 2003
Photogram
31½ x 58¾ inches
Courtesy of the artist

Daniel Lee
Juror No. 6 (Leopard Spirit) (from the series *Judgement*), 1994
Digital inkjet print
53 x 35 inches
Collection of the Museum of Fine Arts, Gift of Marvin and Marlene Maslow, 2003

Daniel Lee
Juror No. 1 (Pig King) (from the series *Judgement*), 1994/2002
Digital C-print
39¼ x 27 inches
Courtesy of the artist

Daniel Lee
Juror No. 7 (Lion King) (from the series *Judgement*), 1994/2002
Digital C-print
39¼ x 27 inches
Courtesy of the artist

Daniel Lee
Juror No. 8 (Carp Spirit) (from the series *Judgement*), 1994/2002
Digital C-print
39¼ x 27 inches
Courtesy of the artist

Daniel Lee
Juror No. 9 (Snake Spirit) (from the series *Judgement*), 1994/2002
Digital C-print
39¼ x 27 inches
Courtesy of the artist

Daniel Lee
Origin, 1999–2003
Digital animation
Five minutes
Courtesy of the artist

The M-M-M Collaborative
(Min Kim Park, Masumi Shibata and Mary Tsiongas, in conjunction with Rex Jung, the MIND Institute)
Threshold, 2006
Interactive touch screen, projection
Dimensions variable
Courtesy of the artists and the MIND Institute, Albuquerque

Patrick Ryoichi Nagatani
Lorentz Transformations, 2006
Faux laboratory diorama
Courtesy of the artist

Exhibition Checklist

Patrick Ryoichi Nagatani
Growth Pulsation (from the series *Chromatherapy*), 2004
Chromogenic print, Type C, Ilfoflex 2000
10 x 18 inches
Courtesy of Andrew Smith Gallery, Santa Fe

Patrick Ryoichi Nagatani
Tonation in Color Charged H_2O (from the series *Chromatherapy*), 2004
Chromogenic print, Type C, Ilfoflex 2000
10 x 18 inches
Courtesy of Andrew Smith Gallery, Santa Fe

Patrick Ryoichi Nagatani
Divinity—Chromatic Conception
(from the series *Chromatherapy*), 2005–2006
Chromogenic print, Type C, Ilfoflex 2000
10 x 18 inches
Courtesy of Andrew Smith Gallery, Santa Fe

Patrick Ryoichi Nagatani
Illumination—Prelude to Dharmakaya Light
(from the series *Chromatherapy*), 2005
Chromogenic print, Type C, Ilfoflex 2000
10 x 18 inches
Courtesy of Andrew Smith Gallery, Santa Fe

Patrick Ryoichi Nagatani
Sliding Filament Theory—Micro Chromatic Healing
(from the series *Chromatherapy*), 2006
Chromogenic print, Type C, Ilfoflex 2000
10 x 18 inches
Courtesy of Andrew Smith Gallery, Santa Fe

Panaiotis et al
The Waters of Life: A Reified Voyage into the Kidney, 2004–2006
Computer simulation, modeling, interactive virtual reality
Dimensions variable
Courtesy of University of New Mexico Health Sciences Center for Telehealth and the Center for High Performance Computing Visualization Laboratory

PRIMARY ARTISTS
Visualization design: Thomas P. Caudell, Andrei Sherstyuk,
    Dale C. Alverson, Stanley M. Saiki, James Holten III,
    Chris Davis, Ken Summers
Music and sound design: Panaiotis
Programming code design and development: James Holten III,
    Panaiotis, Victor M. Vergara, Andrei Sherstyuk,
    Thomas P. Caudell, Ken Summers, Chris Davis

PROJECT CREDITS
University of New Mexico Health Sciences Center:
    Dale C. Alverson, M.D., John Brandt, M.D., Lisa Cerilli, M.D.,
    Kathleen Colleran, M.D., Lee Danielson, Ph.D.,
    Alexis Harris, M.D., Jeffery Norenberg, Pharm.D.,
    Linda Saland, Ph.D., George Shuster, D.N.Sc.,
    Randall Stewart, M.D., Diane Wax, M.P.A., M.B.A.

University of New Mexico School of Engineering:
   Thomas P. Caudell, Ph.D., Chris Davis, James Holten III, Ken Summers, Ph.D., Victor M. Vergara, Ph.D.

University of New Mexico Department of Psychology:
   Timothy Goldsmith, Ph.D., Susan Stevens, MS

University of New Mexico School of Music: Panaiotis, Ph.D.

University of Hawaii John A. Burns School School of Medicine:
   Kathleen Kihmm, Scott Lozonoff, M.D., Jack Lui, Curtis Nakatsu, M.D., Stanley M. Saiki Jr, M.D., Andrei Sherstyuk, Ph.D., Kin Lik ("Alex") Wang

Tripler Army Medical Center, Pacific Telehealth and Technology Hui:
   David Nickles, Stanley M. Saiki Jr, M.D., Daniel Speitel

University of Utah School of Medicine: Martin C. Gregory, M.D.

Gary Schneider
Schneider Family Portrait (from the series *Genetic Self-Portrait*), 2002
Seventeen gelatin silver prints
10 x 8 inches each
Courtesy of the artist and Julie Saul Gallery, New York

Gary Schneider
Retinas (from the series *Genetic Self-Portrait*), 1998
Gelatin silver prints (diptych)
19½ x 20½ inches each
Courtesy of the artist and Julie Saul Gallery, New York

Gary Schneider
Intestinal Flora (from the series *Genetic Self-Portrait*), 1999
Three gelatin silver prints
10 x 8 inches each
Courtesy of the artist and Julie Saul Gallery, New York

Gary Schneider
X and Y Chromosomes (from the series *Genetic Self-Portrait*), 1997
Gelatin silver prints
5⅜ x 3⅞ inches each
Courtesy of the artist and Julie Saul Gallery, New York

Gary Schneider
Sperm (from the series *Genetic Self-Portrait*), 1997
Platinum print
10 x 8 inches
Courtesy of the artist and Julie Saul Gallery, New York

Gail Wight
Ghost, 2004
Plexiglas, electronics
12 x 7 x 8 inches
Courtesy of the artist

Gail Wight
J'ai des papillons noirs tous les jours, 2006
Each of twenty-eight measures 5½ x 5½ x 3½ inches
Silk, electronics, Plexiglas, 2,800 bug pins
Courtesy of the artist

Museum of New Mexico Board of Regents

Karen Durkovich, *President*

Thelma Domenici, *Vice-President*

Mike Stevenson, *Secretary*

Anita Ludovici de Domenico

Maria Estela De Rios

Charlotte Grey Jackson

Margaret Robson

Bev Taylor

New Mexico Department of Cultural Affairs

Stuart A. Ashman, *Secretary*

Troy Fernández, *Deputy Secretary for Administration*

Bergit Salazar, *Deputy Secretary*

Museum of Fine Arts

Marsha C. Bol, Ph.D., *Director*

Mary Jebsen, *Associate Director*

Tim Rodgers, Ph.D., *Chief Curator*

Michael Abatemarco, *Sergeant of Security*

Laura Addison, *Curator of Contemporary Art*

Theresa Garcia, *Administrator/Financial Specialist*

Tim Jag, *Preparator*

Martha Landry, *Special Events Manager*

Dominic Martinez, *Captain of Security*

Christine Mather, *Curator of Collections*

Susan Poorbaugh, *Librarian*

Michelle Roberts, *Chief Registrar*

Merry Scully, *Curator of the Governor's Gallery*

Joan Tafoya, *Registrar*

Joseph Traugott, Ph.D., *Curator of Twentieth-Century Art*

Steve Yates, *Curator of Photography*

Ellen Zieselman, *Curator of Education*